D0578396

WiLDHEART

The DARING Adventures of JOHN MUIR

by Julie Bertagna • illustrated by William Goldsmith

YOSEMITE CONSERVANCY | YOSEMITE NATIONAL PARK

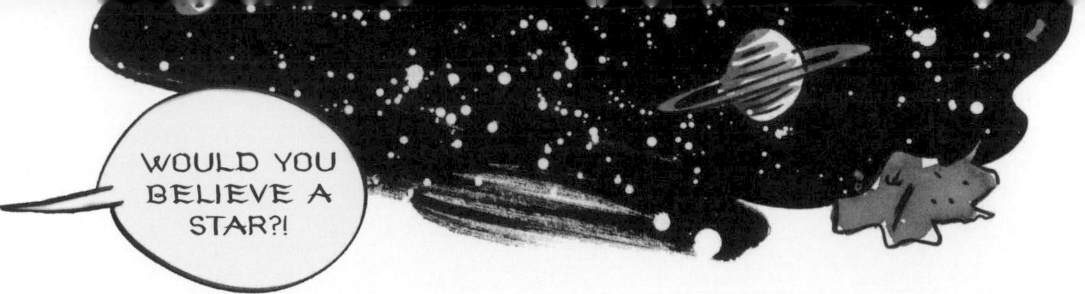

"JOHN MUIR'S PASSION AND COMMITMENT TO PROTECTING AND PRESERVING THE OUTDOORS CONTINUES TO INSPIRE SO MANY OF US TODAY. EVERY SINGLE ONE OF US CAN CONTRIBUTE TO HIS VISION BY PROTECTING AND ENJOYING THE PLACES WE LOVE, AND OUR ACTIONS NOW WILL SPEAK TO ALL FUTURE GENERATIONS."

~ ROBERT HANNA, JOHN MUIR'S GREAT-GREAT-GRANDSON

Text copyright © 2019 by Julie Bertagna
Illustrations copyright © 2019 by William Goldsmith

Published in the United States by Yosemite Conservancy. All rights reserved by Yosemite Conservancy and Scottish Book Trust. No portion of this work may be reproduced or transmitted in any form without the written permission of the publisher, except in the case of brief quotations embodied in critical articles or reviews.

yosemiteconservancy.org

Yosemite Conservancy inspires people to support projects and programs that preserve Yosemite and enrich the visitor experience.

First published in 2014 in Great Britain under the title *John Muir: Earth—Planet, Universe* by Scottish Book Trust in partnership with Creative Scotland and Scottish Natural Heritage. This North American edition is published by arrangement with Scottish Book Trust.

Library of Congress Control Number: 201894806

Cover art by William Goldsmith
Cover design by Melissa Brown
Interior design by Metaphrog, with revisions to the North American edition by Melissa Brown

Trade hardcover ISBN 978-1-930238-94-7 / Library binding ISBN 978-1-930238-93-0

This book was printed on paper from sustainable sources.

Printed in China by Everbest Printing Company through
Four Colour Imports Ltd, Louisville, Kentucky, in September 2018

1 2 3 4 5 6 – 22 21 20 19

Contents

Key Characters

John Muir as a young boy

John Muir as a young man

Stickeen

John Muir as an old man

John Muir's father

John Muir's grandfather

John Muir's mother
John Muir's sister Sarah
John Muir's brother David

John Muir's children, Wanda and Helen

John Muir's wife, Louie

President Theodore Roosevelt

1. The Wild Boy of Dunbar

FOUR TIMES SEVEN IS TWENTY EIGHT

FIVE TIMES SEVEN IS THIRTY FIVE

SIX TIMES SEVEN IS FORTY TWO

I WAS SENT TO SCHOOL WHEN I WAS THREE YEARS OLD.
SEVEN
TIMES
SEVEN
IS...
TWEET TWEET
YAWN

TWEET TWEET!
SKYLARK?

TWEET TWEET!

THWOP!

NO! PAY ATTENTION, BOY!

BUT THE WORLD OUTSIDE
WAS MY REAL SCHOOL.

OATS

O~A~T~S... OATS!
ONE, TWO,
THREE,
FOUR...
OATS
OATS
OATS

PAY ATTENTION, LADDIE!

SMACK!

WATCH OUT, LAD!

BY THE TIME I WAS ELEVEN I KNEW MOST OF THE BIBLE BY HEART.
TIME FOR BIBLE STUDIES, LAD!
AYE, FATHER
8X7
HONOR THY FATHER AND THY MOTHER...
LET MAN HAVE DOMINION OVER THE BIRDS OF THE AIR AND THE FISH OF THE SEA.
MY HEART ACHED TO BE OUTDOORS.
I LOVED EVERYTHING THAT WAS WILD.

?

STOP DREAMING, JOHN!
THWACK!

AYE, FATHER.

I RAN WILD

WHENEVER I COULD.

RACE YE TO THE CASTLE, DAVIE!

I'D CLIMB THE CASTLE LIKE A MOUNTAIN AND SPEND HOURS EXPLORING THE ROCKY SHORE AND FIELDS. THERE, WE WERE AS FREE AS THE WILD CREATURES.

PIRATES!

SHIVER ME TIMBERS, YA SCURVY DOGS!

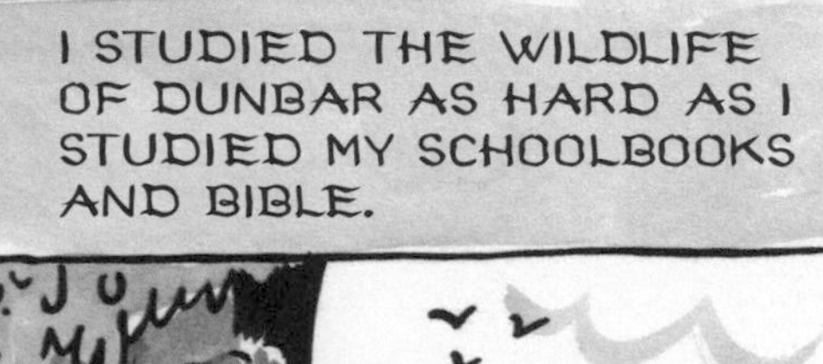
I STUDIED THE WILDLIFE OF DUNBAR AS HARD AS I STUDIED MY SCHOOLBOOKS AND BIBLE.

I COULD TELL A BIRD BY ITS EGGS OR NEST OR SONG.

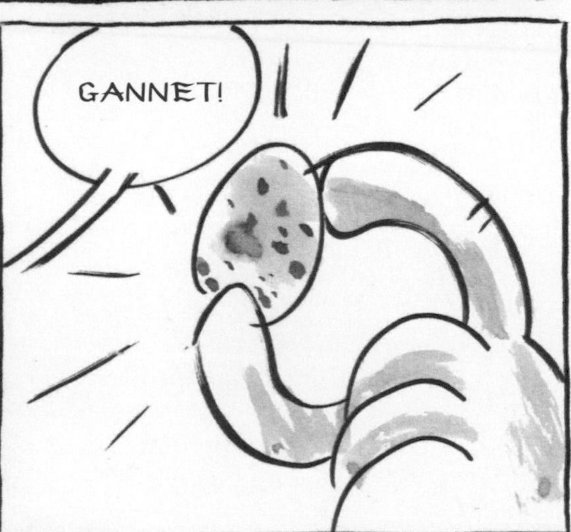
GANNET!

SEAWEED SKIRT!
SEAWEED HAIR!

YIKES! CRAB ATTACK!

WELL, THAT'S US HAD OUR BATHS.

WE LOVED TO WATCH THE WAVES IN AWFUL STORMS, THUNDERING ON THE BLACK HEADLANDS WHEN THE SEA AND SKY CRASHED TOGETHER AS ONE. MOST OF ALL, WE LOVED TO PLAY SCOOTCHERS, A GAME OF DANGEROUS DARES.

GOTCHA!

ONE SCOOTCHER LANDED US IN BIG TROUBLE...
WE'RE AT THE TOP OF THE MOUNTAIN!
HOW DO WE GET DOWN?
SLIDE!
HUP
COME ON! SCOOTCHER!
HELP! I'M STUCK...

I'M FREEZING!

YOOHOO TWIT! SCOOT!

I'LL SAVE YOU!
NOOOO!
IF FATHER CATCHES US HE'LL WARM US WITH HIS BELT!

COME ON, DAVIE! YOU CAN DO IT.

I CAN...

I CAN...
AAAARGH!
GOTCHA!

2. Let's Go West!

ONE DAY MY FATHER GAVE US THE SURPRISE OF OUR LIVES.
TWEET
BAIRNS, PUT AWAY YOUR BOOKS.

WE'RE GOING TO LIVE IN AMERICA!

EH?

NO MORE STUDY! WE'LL LIVE IN BOUNDLESS WOODS FULL OF GOOD THINGS, JOHN. TREES FULL OF SUGAR, GROWING IN GROUND FULL OF GOLD...

YIPPEE!
I'M FREE!

SQUAWK!

WHERE'S
AMERICA?

IS IT NEAR
DUNBAR?

THAT NIGHT, I COULDN'T SLEEP. I WAS SCARED I'D WAKE UP AND FIND IT WAS JUST A DREAM...
ZZZZ

AMERICA!!
NEXT MORNING...
I'LL NO' BE AT SCHOOL TODAY.
I'M AWAY TO AMERICA!
AYE, RIGHT. AND WE'RE AWAY TO THE MOON!

I SET SAIL WITH MY FATHER, BROTHER DAVID, AND SISTER SARAH.

WE'D BUILD A HOUSE BIG ENOUGH FOR US ALL AND THEN MOTHER AND THE YOUNGER CHILDREN WOULD FOLLOW.

LOOK AFTER YOUR WEE BROTHER AND SISTER. NO SCOOTCHERS!
I'LL MISS YOU, MOTHER.

DON'T GREET, ANN. WE'LL BUILD YOU A FINE HOUSE IN AMERICA.
AWAY Y' GO. I'LL MANAGE.

A GOLD COIN FOR A GOLDEN FUTURE.
WOWEE! TA, GRANDPA.

NO' FAIR! HOW COME I DON'T GET ONE?

IT WAS THE MOST EXCITING DAY OF MY LIFE!

ALL ABOARD!

WE SAILED TO AMERICA IN A SHIP CALLED "THE WARREN."
INSIDE THE CABIN...
WHO PACKED ALL THESE BOOKS? THE BIBLE'S ALL WE NEED.
BUNK BEDS!
A SEA VIEW!
JAPAN!
AFRICA!
THE AMAZON!

OUT ON THE VAST WILDERNESS OF OCEAN I SAW THAT EVERYTHING IS HITCHED TO EVERYTHING ELSE IN THE UNIVERSE.
THE WORLD, I'D BEEN TOLD, WAS ALL MADE ESPECIALLY FOR HUMANS ~ BUT WAS IT REALLY?
YUURK!
NUDGE!

SHARK ATTACK!
BEAK ATTACK!
WHALE!
STORM!
STARFISH!
STARS!

AFTER SIX WEEKS AT SEA, FLYING TO OUR FORTUNES ON THE WINGS OF THE WINDS...

LAND AHOY!

IT'S ALL EMPTY!

AN' WILD!

THE OX CART TO WISCONSIN FELT AS BUMPY AS THE SHIP ON A STORMY SEA AS WE TRAVELED ACROSS SWAMPS AND TRACKLESS HILLS.

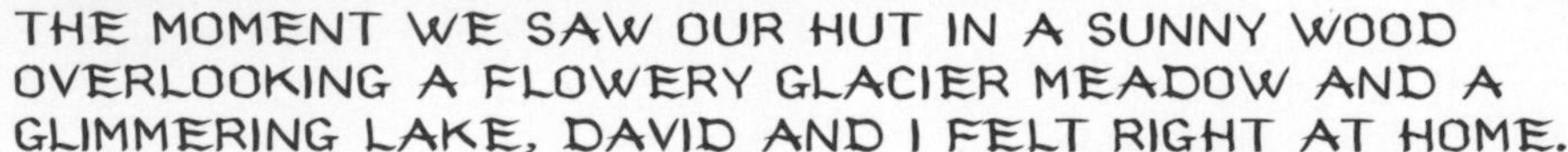

THIS SUDDEN SPLASH INTO A GLORIOUS WILDERNESS MADE US HAPPY AS SKYLARKS.

NATURE STREAMED INTO US, TEACHING US WONDERFUL, WILD LESSONS.

SCREEEECH!
HOW DO WOODPECKERS PECK HOLES SO PERFECT AN' ROUND?

WHY DOES A BLUE JAY HAVE **GREEN** EGGS?

JOHN! DAVID! GET BACK HERE NOW OR I'LL...

FOR A WHILE I WAS FREE TO PLAY IN THE MAGNIFICENT WILDERNESS AND MAKE FRIENDS WITH ALL ITS CREATURES. IT WAS HEAVEN ON EARTH.

TWEET!

OOOH, A NUTHATCH!

HELLO CHICKADEE!

WOWEE, BOBOLINKS!

FOUNTAIN LAKE MEADOW AT TWILIGHT WAS A MAGICAL PLACE, AGLOW WITH MILLIONS OF TWINKLING FAIRY LIGHTS.

LIGHTNING BUGS!

LET'S CATCH 'EM!

3. Genius John in a Brave New World

MY TIME OF FREEDOM WAS SOON OVER. ONCE I TURNED TWELVE, FATHER SAID I WAS OLD ENOUGH TO WORK LIKE A MAN. BUILDING A HOME AND A FARM IN THE WILDERNESS MEANT BACKBREAKING WORK, EVERY DAY, EVERY HOUR OF DAYLIGHT...

CHOPPING...

GRUBBING...

STUMP-DIGGING...

FENCE-BUILDING...

BARN-BUILDING...

HOUSE-
BUILDING...

AND
DIGGING...
DINK
DINK!

LOTS OF
DIGGING.
DINK
DINK!

DINK
DINK!

DAYS AND DAYS OF
DIGGING A WELL...
DINK
DINK!

UNTIL...
DINK-

I HIT A
POCKET OF
POISONOUS
GAS, DEEP
UNDERGROUND.

THEY BROUGHT ME UP IN A
BUCKET, HALF DEAD.

WHIMPER...

THE VERY NEXT DAY I WAS
STRAIGHT BACK AT WORK!
STILL DON'T FEEL WELL.
OCH, YOU'RE FINE!

MY BODY GREW TIRED BUT MY MIND WAS RESTLESS AND HUNGRY.

AFTER SUPPER EACH NIGHT I'D GO TO BED THEN WAKE AT ONE IN THE MORNING

SO I'D HAVE FIVE HUGE HOURS TO READ

BEFORE MY REAL WORK BEGAN AT DAWN.
TWEET TWEET!

THE YEARS FLEW PAST AS MY MIND EXPLODED WITH IDEAS! I BEGAN A FRENZY OF INVENTION: CLOCKS,

AN ANEMOMETER,

A MACHINE TO FEED HORSES,

AND MY AMAZING EARLY-RISING MACHINE! A WAKE-YOU-UP BED!
OOF

I WANTED TO MAKE MY MARK ON THE WORLD. SO I LEFT HOME AND TOOK MY INVENTIONS TO THE WISCONSIN STATE FAIR IN MADISON.
MY TURN!
STOP PUSHING!

CROWDS GATHERED TO SEE ME AND MY MACHINES. SOON I WAS FAMOUS!

HE IS A GENIUS! I CAN NEVER GET THIS BOY OUT OF BED.
YOUNG WIZARD MESMERIZES CROWDS! GENUINE GENIUS!

4. A Terrible Darkness Made Me See

I STILL LOVED NATURE AND BEAUTY, SO ONCE I'D EARNED ENOUGH, I FOLLOWED MY DREAM TO STUDY AT A UNIVERSITY AND LEARN ALL ABOUT THE WORLD.

BUT I LOVED THE BUZZ AND EXCITEMENT OF FACTORIES TOO... WHAT WAS I TO DO WITH MY LIFE?

THEN A TERRIBLE ACCIDENT BLINDED ME ~ AND I FOUND OUT WHAT I REALLY WANTED.
HACK

SLIP!
AAARGH!

MY EYES! EVERYTHING'S DARK!

THE DOCTOR SAID I'D NEVER SEE AGAIN.
I'M SORRY...

THESE WERE TRULY THE DARKEST DAYS OF MY LIFE.

I AM BLIND.
MY LIFE IS
OVER.
BUT AN EYE SPECIALIST SAID
IF I RESTED FOR LONG WEEKS
IN A DARK ROOM...

TWEET

TWEET

TWEET
TWEET

WILL I BE
TRAPPED IN
DARKNESS
FOREVER?

THEN, ONE DAY
TWEET
TWEET

I CAN HEAR THE WIND AND THE BIRDS, BUT ALL IS DARK.

WAIT... WHAT'S THAT STRANGE LIGHT?

TUG

TUG

TWEET!
AND NOW I KNEW EXACTLY WHAT I WANTED TO DO.

I'M CURED! I CAN SEE! AND I WANT TO BE OUT IN THE WILD BEAUTY OF THE WORLD FOR THE REST OF MY DAYS!

5. I Will Walk One Thousand Miles

I WAS LIKE A MAN LOST IN THE DESERT WHO FINDS AN OASIS AND MUST DRINK.
YAWN
STRETCH

BUT I WAS THIRSTY FOR THE WHOLE WORLD!
SPLISH!

I WANTED TO JUMP INTO IT AND IMMERSE MYSELF IN ITS WONDERS.

LACE!
TIE!

SO I WENT ON A THOUSAND-MILE WALK TO DRINK IN THE GRAND SHOW OF NATURE.

I'LL FOLLOW MY NOSE. THE WIND IS FULL OF HAUNTING, UNKNOWN SCENTS...

STRANGE ANIMAL CRIES...
OOOOEE

BEAUTIFUL NEW PLANTS AND FLOWERS ...

IN TENNESSEE I FOUND SOMETHING I'D DREAMED OF FROM BOYHOOD,

BUT HAD NEVER SEEN BEFORE.
A MOUNTAIN! MY FIRST MOUNTAIN!

I HAD MANY ADVENTURES ALONG THE WAY...
HANDS UP AN' GIMME YOUR MONEY!

I'VE NO MONEY, JUST...
CLATTER

THESE.
BURNS' POEMS
SOAP

IS THAT IT?!
BURNS' POEMS

HAVE IT BACK, THEN.

WITH ALL MY MONEY GONE, I BUILT A HUT OUT OF BRANCHES IN SAVANNAH, GEORGIA.

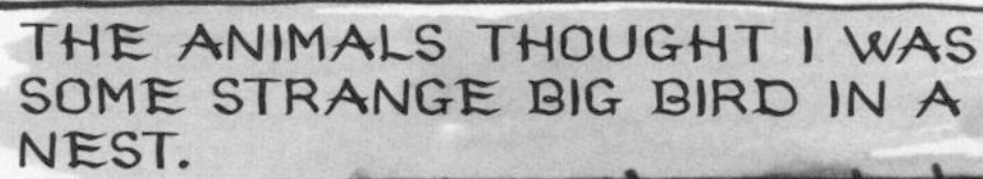
THE ANIMALS THOUGHT I WAS SOME STRANGE BIG BIRD IN A NEST.

?
?

BURNS' POEMS

LOOKS A BIT NUTTY TO ME.

ONCE MY FAMILY HAD SENT MY SAVINGS AND I HAD MONEY IN MY POCKET AGAIN, I PLANNED TO SAIL FOR SOUTH AMERICA.

BUT IN FLORIDA I
CAUGHT MALARIA.

UURGH...
BZZZ

NIP!

OUCH!

THE PURE MOUNTAIN AIR OF THE YOSEMITE VALLEY WILL CURE ME. THAT'S WHERE I MUST GO!
MAP

I HAD SEEN SO MUCH ON MY THOUSAND-MILE JOURNEY, AND I WOULD TRAVEL MANY THOUSANDS OF MILES MORE IN MY LIFE.

AFTER A VISIT
TO HAVANA, CUBA,
I SAILED FROM
NEW YORK TO
CALIFORNIA.

6. The Mountains Are Calling Me!

A VAST VALLEY OF FLOWERS STRETCHED BEFORE ME LIKE A LAKE FULL OF RAINBOWS. BEYOND THAT WERE THE GIANT PEAKS OF THE MOUNTAINS.

I CALLED THESE MOUNTAINS THE RANGE OF LIGHT.
SO MUCH BEAUTY! THE MOUNTAINS GLOW IN THE SUN AS IF THEY'RE MADE OF LIGHT.

THEN I SET OFF FOR THE YOSEMITE VALLEY I SO LONGED TO SEE.

YOSEMITE
YO~ SEY~ M...?
"YO~SEM~IT~EE" IS HOW YOU SAY IT!
YOSEMITE

THE WINDS OF THE YOSEMITE VALLEY BLEW THEIR FRESHNESS INTO ME.

THE STORMS GAVE ME THEIR ENERGY.

SOON MY CARES FELL AWAY LIKE AUTUMN LEAVES.

I FEEL WELL AGAIN!

THE HOPE OF THE WORLD LIES IN THE GREAT UNSPOILED WILDERNESS.

I TOOK A JOB AS A SHEPHERD SO THAT I COULD STUDY THE YOSEMITE. I WANTED TO WRITE ABOUT IT AND SHOW NATURE'S LOVELINESS AND WONDER TO THE WORLD.

CLIMB THE MOUNTAINS AND NATURE'S PEACE WILL FLOW INTO YOU AS SUNSHINE FLOWS INTO TREES.

WE'RE FLYING THROUGH SPACE WITH THE OTHER STARS, ALL SHINING TOGETHER AS ONE.

THE WHOLE UNIVERSE IS AN INFINITE STORM OF BEAUTY!

ONE MOUNTAIN TOOK ME SEVEN HOURS TO CLIMB...
AND A MINUTE TO COME BACK DOWN AGAIN...
AVALANCHE! LET THE POWER OF NATURE SWEEP YOU AWAAAAY!

I DISCOVERED THAT THE CLEAREST WAY INTO THE UNIVERSE IS THROUGH A FOREST WILDERNESS.

I NEVER SAW AN UNHAPPY TREE.

TREES TRAVEL AS FAR AS WE DO, WANDERING IN ALL DIRECTIONS,

WITH EVERY WIND, TRAVELING WITH US AROUND THE SUN,

TWO MILLION
MILES A DAY!

AND THROUGH SPACE,
HEAVEN KNOWS HOW
FAST AND FAR...
OOOH!
TWIT!

BETWEEN EVERY
TWO PINES THERE IS
A DOOR LEADING TO
A NEW WAY OF LIFE.

THIS
IS THE
LIFE!
DINNER!
DINNER!

7. Companions in Peril

AFTER YEARS IN THE MOUNTAINS, I CAME BACK TO THE HUMAN WORLD. I MARRIED AND BECAME A FRUIT FARMER.

SOON LOUIE AND I HAD TWO DAUGHTERS, WANDA AND HELEN.

DAISY, BUTTERCUP, FORGET-ME-NOT, DANDELION...
DADDY LION?
DO WE NEED TO LEARN THE NAMES OF ALL THE FLOWERS?

YOU WOULDN'T LIKE IT IF PEOPLE DIDN'T KNOW YOUR NAMES!

BUT AFTER TEN YEARS AS A FARMER, I FELT LIKE A CAGED LION AND LONGED FOR THE WILDERNESS AGAIN.

LOUIE SOLD PART OF THE RANCH SO THAT I HAD MONEY FOR MY NATURE WORK.

HERE'S FRUIT FOR THE JOURNEY, PAPA!
PLUM, APPLE, PEAR.

OFF YOU GO THEN! WE'LL MANAGE.

YOU'RE THE BEST.

HIGH IN THE SIERRA NEVADA, I'D DISCOVERED A GLACIER AND SAW HOW THESE VAST, ANCIENT RIVERS OF ICE MUST HAVE CARVED OUT THE GREAT VALLEY.

I WANTED TO PROVE MY THEORY, SO I WENT TO SEE THE LIVING GLACIERS OF ALASKA.

MY COMPANION WAS A DOG CALLED STICKEEN.

TOGETHER WE EXPLORED GLACIER BAY, AN UNKNOWN INLET SURROUNDED BY MOUNTAINS AND FULL OF ICEBERGS.

I STUDIED THE DEEP CREVASSES AND THE POWERFUL MOVEMENT OF THE GLACIER ICE AS IT MOVED SLOWLY, SLOWLY THROUGH THE MOUNTAINS, CARVING A PATH THROUGH THE STONY LANDSCAPE.
SCRIBBLE
WOOF!
BE CAREFUL, STICKEEN!

EVERYTHING
IS FLOWING~
GOING
SOMEWHERE,
ANIMALS

AND SO~CALLED
LIFELESS ROCKS

AS WELL AS WATER.

GRAND GLACIERS

AND AVALANCHES,

MAJESTIC FLOODS CARRYING
MINERALS AND ROCKS,

AIR CARRYING PLANT LEAVES AND SPORES...

WHILE THE STARS GO STREAMING THROUGH SPACE PULSING ON AND ON FOREVER LIKE BLOOD GLOBULES IN NATURE'S WARM HEART.

THE SUN FELL. THE STORMY DARKNESS SWIRLED AROUND US AS WE TRIED TO HURRY HOMEWARD.

WE STRUGGLED ACROSS A MAZE OF JUMBLED ICE AND DEATHLY CREVASSES.

WE'RE MAROONED ON AN ISLAND OF ICE!
WHIMPER...
THEN WE CAME TO A CHASM SO WIDE AND DEEP WE COULDN'T JUMP IT. AND WE COULDN'T GO BACK. THE BANK BEHIND WAS TOO HIGH.

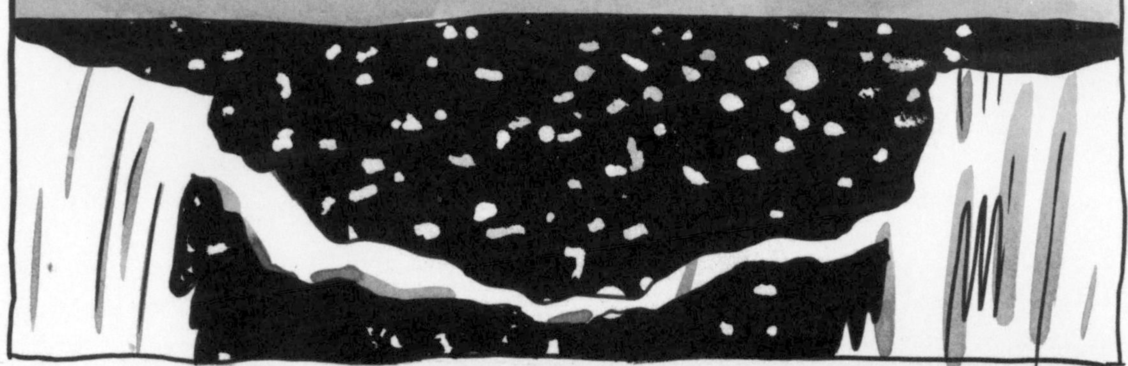
A LONG, THIN BRIDGE OF ICE WAS OUR ONLY HOPE ACROSS THE CHASM. BUT IT SAGGED IN THE MIDDLE, MELTING.

DINK
DINK!
WE'D HAVE TO CLIMB DOWN TO REACH THE BRIDGE, THEN CLIMB UP A SHEER CLIFF OF ICE ON THE OTHER SIDE. COULD WE DO IT?

FOLLOW ME, STICKEEN. SEE, I'VE CUT STEPS IN THE ICE FOR US. DON'T BE SCARED.
I WAS TERRIFIED!

YOU CRAZY HUMAN!

MY FINGERS FROZE AS I CLIMBED. IT WAS PERILOUS, THE DEADLIEST THING I'VE EVER DONE.

MADE IT!

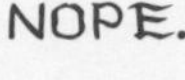
NOPE.

WHEN STICKEEN SAW THERE WAS NO OTHER WAY, HE VENTURED ONTO THE BRIDGE.

I HELD MY BREATH AS THE ICY WINDS BLASTED HIM, WILLING HIM NOT TO FALL. DOGS ARE NOT GOOD CLIMBERS...

COULD HE MAKE IT TO THE OTHER SIDE?
SKID!

AWOOOOO!

WOOOOOO!

YOU CAN DO IT, BOY!

COME ON, DAVIE! YOU CAN DO IT.

I'LL SAVE YOU!

I CAN, I CAN

COME ON, STICKEEN! YOU CAN DO IT!

ONE, TWO,
THREEEEEE...!
HOP
HOP
HOP
HOP
IN A GREAT WILD BOUND,
STICKEEN LEAPT UP
THE STEPS, FAST AS
LIGHTNING ~ AND LANDED
SAFELY BESIDE ME!

THAT AMAZING DOG RACED ABOUT LIKE A MAD CREATURE, BARKING AND HOWLING AND ROLLING IN THE SNOW, ALMOST BURSTING WITH JOY TO BE ALIVE.

FROM THAT DAY, WE WERE SOUL MATES.

STICKEEN HAD TAUGHT ME ONE OF THE MOST IMPORTANT LESSONS OF MY LIFE ~

THAT HUMANS AND ANIMALS ARE TRUE COMPANIONS ON THIS EARTH,

HAHAHA! SEE ME? SEE THAT?

8. The Wild Man Who Changed the World

MY EXPERIENCE IN THE ALASKAN GLACIERS REVEALED AMAZING THINGS TO ME.
THANKS, STICKEEN!

ANIMALS WERE NOT, AS PEOPLE OF MY TIME BELIEVED, PUT ON EARTH TO SERVE THE HUMAN RACE.
PFZZ

I MEAN, WHAT GOOD IS AN ALLIGATOR TO ANYONE?

FISH SUPPER!

AND I WAS CERTAIN NOW THAT ANCIENT GLACIERS REALLY HAD CARVED OUT THE YOSEMITE VALLEY AND THE SIERRA, LIKE GIANT ICE-CUTTERS.
SILLY SHEEP-HERDER!
GLACIERS? WHAT NONSENSE!
TWADDLE!

NO ONE ELSE BELIEVED THIS NEW THEORY, SO I WENT BACK TO MY MOUNTAINS, DETERMINED TO FIND PROOF...

THESE GROOVES AND MARKINGS SHOW THAT RIVERS OF ICE GOUGED OUT VAST PATHS AS THEY MOVED THROUGH THE BEDROCK.

BY JOVE, HE'S RIGHT!
THE GLACIERS OF THE LAST ICE AGE MADE THIS VALLEY!

LET CHILDREN WALK WITH NATURE, IN WOODS AND MEADOWS, PLAINS AND MOUNTAINS. OUR BODIES WERE MADE TO THRIVE WHERE PURE AIR IS FOUND.

SO I BEGAN MY CAMPAIGN TO PROTECT AMERICA'S FORESTS.

A FOREST OF 3,000-YEAR-OLD GIANTS - ALL GONE IN A MOMENT.

OH, THE WRONGS DONE TO TREES!
THE PEOPLE LISTENED AND THEY FORCED THE POLITICIANS TO ACT.
SAVE OUR FORESTS OR WE'LL HAVE NONE LEFT!
DEFEND THE TREES!

I WANTED THE AMERICAN PEOPLE TO LOVE AND PROTECT NATURE, ESPECIALLY THE BEAUTIFUL HIGHLANDS OF CALIFORNIA. SO I CO-FOUNDED THE SIERRA CLUB. I WAS PRESIDENT FOR 22 YEARS.

I ALSO HELPED CREATE NATIONAL PARKS IN MANY WONDERFUL WILD PLACES LIKE YOSEMITE
YOSEMITE NATIONAL PARK

AND THE GRAND CANYON.

FRIENDS REUNITED IN THE WILD!

DO SOMETHING FOR THE WILDERNESS AND MAKE THE MOUNTAINS GLAD!

THEN I MET ANOTHER PRESIDENT...THE PRESIDENT OF THE UNITED STATES OF AMERICA.
MY TALKS AND WRITINGS HAD MADE ME A FAMOUS MAN.
LET'S GO CAMPING IN THE WILD!
YES, PRESIDENT ROOSEVELT.
BUT MR. PRESIDENT, WE HAVE A GRAND DINNER READY IN YOUR HONOR...
SORRY, TOO BUSY! TAKE ME TO YOSEMITE, JOHN.

FOUR DAYS WE CAMPED IN THE YOSEMITE. ME, JOHN MUIR FROM DUNBAR,

AND PRESIDENT THEODORE ROOSEVELT! WE CLIMBED TO GLACIER POINT

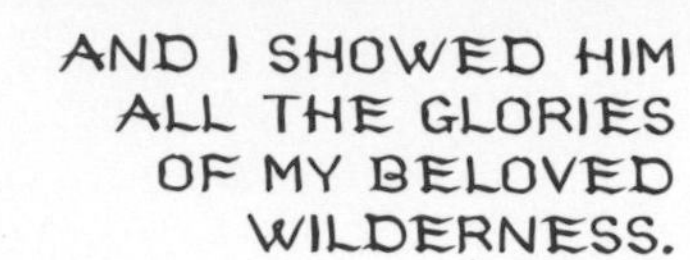
AND I SHOWED HIM ALL THE GLORIES OF MY BELOVED WILDERNESS.

WE TALKED ABOUT PROTECTING THE FORESTS AND NATURE.

I'LL PRESERVE 148 MILLION ACRES OF FOREST AND CREATE MANY NEW NATIONAL PARKS!

HURRAH! WE MUSTN'T WASTE THE WILDERNESS. IT'S A NECESSITY~ NOT JUST AS FOUNTAINS OF TIMBER BUT AS FOUNTAINS OF LIFE ITSELF.

TREES ARE THE LUNGS OF THE WORLD ~
OUR FRIENDS AND PROTECTORS.

NO MATTER HOW TIMES
CHANGE, WE MUST LOOK
AFTER OUR FORESTS.

IF WE KILL THE TREES,
WHAT WILL HAPPEN TO US?

9. As the Round Earth Rolls...

MY LIFE HAS BEEN ONE BIG ADVENTURE.

I HAD A DREAM AND I WORKED AND WORKED AT IT.

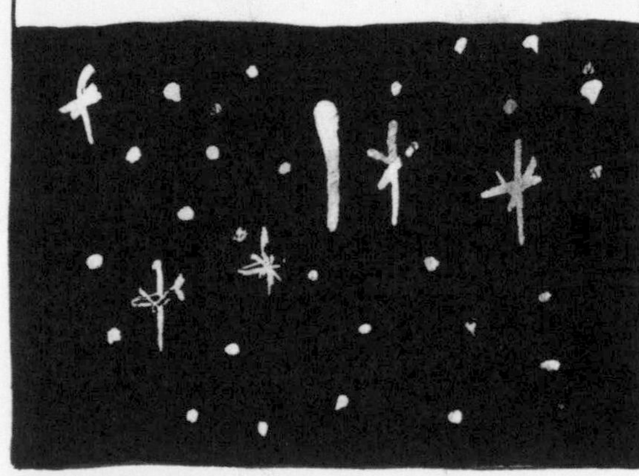
I HITCHED MY WAGON TO THAT STAR.

AGED 65, I WENT ON A WORLD TOUR TO EVERY CONTINENT EXCEPT ANTARCTICA.

I CROSSED OCEANS TO EXPLORE THE GREAT RIVERS AND FORESTS OF THE EARTH.

AT 73, I MADE A 40,000-MILE TREK TO SOUTH AMERICA AND AFRICA.

I'LL FOLLOW MY
HEART TILL THE END
OF MY DAYS, AND SEEK
THE WILD SPIRIT OF
THE WORLD.

I'D ALSO TRAVELED
BACK TO MY ROOTS, TO
DUNBAR, WHERE MY LOVE
OF ADVENTURE AND
CURIOSITY ABOUT THE
WORLD BEGAN.

I WAS MUCH CHANGED FROM THE GANGLY BOY WHO'D
ONCE CLIMBED THE CRAGS OF THE CASTLE.

BUT EVERYWHERE I
WENT, PEOPLE RUSHED
OUT TO GREET ME.

CHILDREN FOLLOWED
AS IF I WERE THE
PIED PIPER!

MY OLD HOME!

I REMEMBERED OUR SCOOTCHER...

THAT'S JOHN MUIR. I KNEW HIM WHEN HE WAS A SKINNY-MALINKY LADDIE. HE'S WORLD FAMOUS NOW!

IS HE? IS HE RICH?

GO TO THE WILD PLACES AND LISTEN TO WHAT THEY SAY TO YOU.

HEAR THE MUSIC OF THE WIND.

WATCH THE PERFECT FLIGHT OF A BIRD.

FEEL THE LIFE AROUND YOU AND INSIDE YOU.

BE JOYFULLY LOOSE AND LOST IN THE WORLD!

EVERYBODY NEEDS BEAUTY AS WELL AS BREAD...

AND PLACES TO PLAY IN!

MY LEGACY
WOULD LIVE ON
LONG AFTER I
WAS GONE.

MY FACE WOULD
BE ON STAMPS
AND COINS.

ALL KINDS OF
THINGS WOULD BE
NAMED AFTER ME:

A TINY
PLANET

AN ALPINE
RABBIT

A GLACIER

A BUTTERFLY

A ROSE

A BIRD

A TREE

MOUNTAINS

PARKS

SCHOOLS

AND TRAILS.
EVEN...

A MILLIPEDE?
HELLO, I'M JOHN MUIR!

STILL THE WILDERNESS CALLS ME, THOUGH I'M TOO OLD FOR ADVENTURES NOW.

I SIT BY MY FIRE AND SAIL OUT ON A SEA OF MEMORIES, RELIVING MY WONDERFUL LIFE.

ONE MEMORY KEEPS RUSHING BACK
CRACKLE

LIKE A WAVE TO THE SHORE,
BARK

CLEARER AND STRONGER THAN ALL THE REST.

THAT DAY ON THE ALASKAN GLACIER STILL ROARS IN MY MIND.

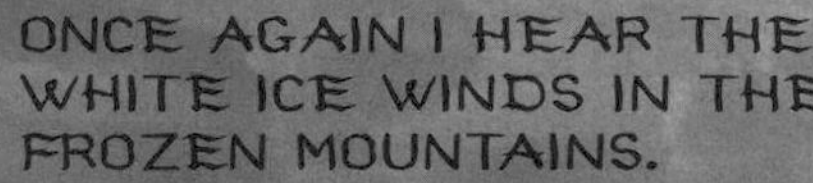
ONCE AGAIN I HEAR THE WHITE ICE WINDS IN THE FROZEN MOUNTAINS.

STICKEEN IS AT MY SIDE, AND TOGETHER WE BATTLE THE STORM-BLAST

TO CROSS THE DEATHLY CREVASSE.

IT'S ALWAYS SUNRISE SOMEWHERE.
ETERNAL SUNRISE, ETERNAL DAWN, AS THE ROUND EARTH ROLLS...
WHAT I DISCOVERED ON THE GLACIER HAS STAYED WITH ME ALWAYS.

WE ARE PART OF NATURE, AND ITS WILD HEART IS PART OF US.
THE HUMAN CREATURES OF THIS PLANET ARE CONNECTED
TO ALL OUR EARTH-BORN COMPANIONS.

AND I SAW HOW A SNOWFLAKE COULD MOVE A MOUNTAIN. IF MILLIONS OF SNOWFLAKES CAN BOND TOGETHER
TO CREATE THE SLOW~MOVING POWER OF A GLACIER ~
STRONG ENOUGH TO MOVE MOUNTAINS
AND CARVE OUT VALLEYS ~

THEN MILLIONS OF SMALL ACTIONS BY EACH HUMAN BEING,

ALL THE LITTLE THINGS WE DO EVERY DAY,

CAN SURELY ADD UP TO VAST CHANGES OVER TIME.

TODAY, THE WORLD IS STILL INSPIRED BY THE LIFE AND VOICE OF ONE MAN IN THE WILDERNESS: A SON OF SCOTLAND, AN ADOPTED SON OF AMERICA, A PIONEER PLANET-KEEPER, AND THE FIRST MODERN ENVIRONMENTALIST.

JOHN MUIR CALLS OUT TO US ALL TO LOOK BEYOND OUR EVERYDAY LIVES AND ENJOY THE WONDERS OF NATURE ~ TO SEE OURSELVES AS CITIZENS OF THE UNIVERSE AND CARETAKERS OF THE EARTH FOR FUTURE GENERATIONS.

John Muir's Chronology and Legacy

1838	April 21: John Muir born in Dunbar, Scotland.
1841	Begins school.
1849	Immigrates to Wisconsin at the age of eleven.
1860	Leaves home (22 years of age) to exhibit his inventions at the State Agriculture Fair in Madison, Wisconsin.
1861–2	Attends the University of Wisconsin.
1867	Suffers an eye injury; begins one-thousand-mile walk to Florida and then to Cuba.
1868	Sails from New York to San Francisco, then walks across California to Yosemite. Falls in love with the Sierra Nevada and Yosemite. After all his travels, John settles in Yosemite for several years. Also begins exploration for signs of glaciers.
1871	Publishes articles in leading magazines and becomes a writer.
1874	Begins his tree-saving campaign and puts forward his glaciation theory for the formation of Yosemite Valley.
1879–90	Goes on many expeditions to Alaska.
1880	April 14: John marries Louie Wanda Strentzel. They have two daughters and settle at their ranch.
1890	Because of his many articles he begins the movement on wilderness conservation. Yosemite is established as a national park.
1892	Helps to create the Sierra Club and is the president until his death.
1893	Begins a European trip.
1901	Publishes his book *Our National Parks.*

1903 President Theodore Roosevelt journeys to visit Yosemite in mid-May. He spends his time there with John Muir, camping, hiking, and riding horseback.

1903–5 John's world tour.

1907–13 Fights to save Hetch Hetchy Valley. Many of John's books are published.

1908 Muir Woods National Monument established in Mill Valley, California.

1911–12 John goes on trips to South America and Africa (73 years old).

1914 December 24: John Muir dies of pneumonia in a Los Angeles hospital.

1919 John's earlier activism leads to the establishment of Grand Canyon National Park.

1938 John Muir Trail completed—a 212-mile Sierra trail passing through Yosemite, Sequoia, and Kings Canyon National Parks.

1964 John Muir National Historic Site established at the Muir home in Martinez, California.

1976 John Muir Country Park designated at Dunbar, Scotland.

1981 John Muir's Birthplace museum opens in Dunbar, Scotland.

1983 John Muir Trust is founded in Scotland to conserve wild land. It later purchases Scottish land and mountains, including Schiehallion and Ben Nevis.

1988 April 21: John Muir Day established in California.

2005 John Muir appears with Half Dome and a California condor on the 2005 California State Quarter.

2006 Los Angeles-based amateur astronomer R. E. Jones officially names a tiny planet he discovered in 2004 "Johnmuir."

Sources: John Muir National Historic Site, John Muir Trust, and John Muir Exhibit at sierraclub.org

Glossary

A

anemometer A device used for measuring wind speed.

avalanche Large amounts of snow and ice that have become dislodged and fall rapidly down a mountainside.

B

bairn A Scots word for child or baby.

bobolink A bird, from the same family as the blackbird, that nests in North America and winters in South America. Bobolinks have white or yellow feathers on their heads.

C

chasm A deep, steep-sided opening in the earth's surface.

chickadee A small songbird, from the same family as a titmouse, that lives in North America and Canada.

continent A huge, continuous expanse of land, which can contain many countries. The world's continents are Europe, Asia, Africa, North America, South America, Australia, and Antarctica.

crevasse A deep, open crack, usually in a glacier.

G

gangly Tall, thin, and not very graceful.

gannet A large sea bird, a bit like a gull. Gannets are black and white and live near coasts in many parts of the world, including North America.

glacier A large body of ice that moves, made from layers and layers of compacted snow. Glaciers form on land and move very, very slowly over it.

glaikit A Scots word for stupid, foolish, or absentminded.

	greet	A Scots word for cry or weep.
I	**inlet**	A narrow passage of water at a spot where a coastline indents.
K	**kin**	Another word for family.
L	**legacy**	What a person leaves behind after they die; what others remember about them.
M	**malaria**	A disease caused by a parasite transferred through mosquito bites and that is characterized by chills and fever.
N	**nuthatch**	A small bird, often brightly colored, that lives in woodlands. Nuthatches are mostly found in North America, Asia, and Europe.
P	**perilous**	Dangerous or risky.
R	**rascal**	An affectionate term for a mischievous person.
S	**scootchers**	John Muir's made-up word for adventures; a game of dares.
	scurvy	A disease that results from not getting enough vitamin C. Once upon a time it was common in sailors. "Scurvy dogs" is a seafaring term for sailors who are considered bad, dirty, or diseased.
	skinnymalinky	A Scots word for a very thin person.
	skylark	A small brown-and-white songbird with a crest on its head. Skylarks live mainly in Europe, Asia, and North Africa.
T	**ta**	A Scots word for thank you.

Sources of Inspiration

Excerpted or adapted from *John of the Mountains: The Unpublished Journals of John Muir*, edited by Linnie Marsh Wolfe (Madison: University of Wisconsin Press, 1938). Copyright © 1938 by Wanda Muir Hanna. John Muir Papers, University of the Pacific © 1984 Muir-Hanna Trust.

p. 73: "The hope of the world...," p. 317.
p. 76: "The clearest way into the Universe...," p. 313.
p. 76: "I never saw an unhappy tree...," p. 313.
p. 100: "Our bodies were made to thrive...," p. 191.
p. 101: "The wrongs done to trees...," p. 429.
p. 116: "It's always sunrise somewhere...," p. 438.

Adapted from John Muir to Sarah Galloway letter, September 3, 1873. John Muir Papers, University of the Pacific © 1984 Muir-Hanna Trust.

p. 70: "The mountains are calling me!"

Adapted from a handwritten margin note in Muir's copy of *The Prose Works of Ralph Waldo Emerson,* vol. 1. Beinecke Rare Book and Manuscript Library, Yale University.

p. 77: "Between every two pines. . . ."

John Muir, *The Mountains of California* (New York: The Century Co., 1894).

p. 111: "Be joyfully loose and lost...," p. 72.

John Muir, *My First Summer in the Sierra* (Boston: Houghton Mifflin, 1911).

p. 34: "Everything is hitched...," p. 216.
p. 84: "Everything is flowing...," p. 316.

John Muir, *Our National Parks* (Boston: Houghton Mifflin, 1901).

p. 74: "Climb the mountains...," p. 56.
p. 104: "We mustn't waste the wilderness...," p. 2.

John Muir, *Story of My Boyhood and Youth* (Boston: Houghton Mifflin, 1913).
p. 14: "I loved everything that was wild," p. 1.
p. 21: "We loved to watch the waves...," p. 2.
p. 28: "No more study...," p. 53.
p. 40: "How do woodpeckers...," p. 65.
p. 40: "This sudden splash...," p. 63.

John Muir, *A Thousand-Mile Walk to the Gulf* (Boston: Houghton Mifflin, 1916).
p. 34: "The world, I'd been told...," p. 136.
p. 100: "Let children walk with nature...," p. 70.

John Muir, *Travels in Alaska* (Boston: Houghton Mifflin, 1915).
p. 74: "We're flying through space...," p. 6.

John Muir, *The Yosemite* (New York: The Century Co., 1912).
p. 111: "Everybody needs beauty...," p. 256.

Sierra Club Bulletin, vol. 10, no. 2 (1917).
p. 102: "Do something for the wilderness...," p. 138.

Acknowledgments

Sincere thanks to the wonderful Marc Lambert, Koren Calder, Philippa Cochrane, and all at Scottish Book Trust who made this book possible and were vital in its development. Thanks also to Creative Scotland, Scottish Natural Heritage, the John Muir Trust, and students and educators around Scotland for fantastic input, help, and support. And many thanks to Yosemite Conservancy and Nicole Geiger for giving our story wings.

About the Author and Illustrator

JULIE BERTAGNA is an award-winning Scottish author whose acclaimed books and short stories for children and young adults have been published around the world. *Exodus,* a modern classic, has won various awards, including a Friends of the Earth Eco Prize for Creativity and a Santa Monica Public Library Green Prize for Literature. Five of her young-adult novels have been nominated for the United Kingdom's prestigious Carnegie Prize and other major short lists, such as the Whitbread Children's Book of the Year and the Booktrust Teenage Prize. Awards in her native Scotland include the Children's Book of the Year and the Catalyst Book Prize. For more about Julie, visit juliebertagna.com and @JulieBertagna on Twitter.

WILLIAM GOLDSMITH is a British writer and illustrator based in Moscow, where he teaches illustration at the British Higher School of Art and Design. His work has appeared in a range of publications and exhibitions in the United Kingdom and overseas. His other graphic novels are *The Bind* and *Vignettes of Ystov,* published by Jonathan Cape. He is currently working on a middle-grade chapter book. For more about William, visit williamgoldsmith.co.uk.

Parks Are for You

John Muir was America's most famous and influential naturalist and preservationist. He worked tirelessly to save wild places, and his message reached people far and wide, including lawmakers. Muir's writings in newspapers, magazines, books, and letters helped convince the U.S. government to protect Yosemite, the Grand Canyon, Mount Rainier, many of California's ancient sequoia groves, and other areas as national parks and monuments.

People today visit parks and recreation areas that bear John Muir's name, including the John Muir National Historic Site at his family's former home in Martinez, California; the Muir Woods National Monument, home to colossal coast redwoods near San Francisco; the John Muir Trail, a 210-mile route in California's Sierra Nevada; the 21-mile John Muir Trail in Tennessee's Cherokee National Forest; and the quarter-mile John Muir Ecological Park trail, in Yulee, Florida. If you visit Scotland, you can hike and bike the John Muir Way from coast to coast, right through the center of the country. Really, almost any park or wild land you visit owes a debt to Muir, who never stopped fighting on the side of nature.

Honor John Muir by visiting a park near you! Many parks are accessible via public transportation, some will require having a car, or one may be within walking distance. Fees to enter parks will vary. Check to see if the park you wish to visit is having a fee-free day. Be sure to bring a backpack stocked with snacks, water (at least one liter for the day), and lunch for a picnic. Also bring sun protection, extra clothing, and athletic or hiking shoes. It's also a good idea to have a map of where you are going since a cell phone or GPS device may not always work. Don't forget to practice "leave no trace" principles while you are there: throw away all waste, including crumbs, in proper receptacles, or better yet, dispose of your trash back at home.

The most important thing is to have fun exploring a protected natural place that belongs to everyone—including you. Only outdoors can we realize that so much of what John Muir experienced and shared long ago is just as true today: people need wild places, and wild places need people to protect them.